Spotlight on Georgia Performance Standards

HSP Georgia
Science

Interactive Text

Harcourt
SCHOOL PUBLISHERS

Visit *The Learning Site!*
www.harcourtschool.com

Printed in the United States of America

ISBN 978-0-15-378393-7
ISBN 0-15-378393-1

16 17 18 0914 16 15 14 13   4500425348

Chapter 2 Patterns in the Sky

Georgia Performance Standards in This Chapter

S2E1 Students will understand that stars have different sizes, brightness, and patterns.

S2E1a Describe the physical attributes of stars—size, brightness, and patterns.

S2E2 Students will investigate the position of sun and moon throughout the year to show patterns throughout the year.

S2E2a Investigate the position of the sun in relation to a fixed object on earth at various times of the day.

S2E2b Determine how the shadows change through the day by making a shadow stick or using a sundial.

S2E2c Relate the length of the day and night to the change in seasons (for example: Days are longer than the night in the summer.).

S2E2d Use observations and charts to record the shape of the moon for a period of time.

Chapter 3 Changes to the Earth

Georgia Performance Standards in This Chapter

S2E3 Students will observe and record changes in their surroundings and infer the causes of the changes.

S2E3a Recognize effects that occur in a specific area caused by weather, plants, animals, and/or people.

Chapter 4 States of Matter

Georgia Performance Standards in This Chapter

S2P1 Students will investigate the properties of matter and changes that occur in objects.

S2P1a Identify the three common states of matter as solid, liquid, or gas.

Chapter 5 Changes in Matter

Georgia Performance Standards in This Chapter

S2P1 Students will investigate the properties of matter and changes that occur in objects.

S2P1a Identify the three common states of matter as solid, liquid, or gas.

S2P1b Investigate changes in objects by tearing, dissolving, melting, squeezing, etc.

Chapter 6 Energy

Georgia Performance Standards in This Chapter

S2P2 Students will identify sources of energy and how the energy is used.

Chapter 7 Motion

Georgia Performance Standards in This Chapter

S2P2   Students will identify sources of energy and how the energy is used.

     S2P2b   Describe how light, heat, and motion energy are used.

S2P3   Students will demonstrate changes in speed and direction using pushes and pulls.

     S2P3a   Demonstrate how pushing and pulling an object affects the motion of the object.

     S2P3b   Demonstrate the effects of changes of speed on an object.

Chapter 8 Life Cycles

Georgia Performance Standards in This Chapter

S2L1   Students will investigate the life cycles of different living organisms.

     S2L1a   Determine the sequence of the life cycle of common animals in your area: a mammal such as a cat or dog or classroom pet, a bird such as a chicken, an amphibian such as a frog, and an insect such as a butterfly.

     S2L1b   Relate seasonal changes to observations of how a tree changes throughout a school year.

     S2L1c   Investigate the life cycle of a plant by growing a plant from a seed and by recording changes over a period of time.

     S2L1d   Identify fungi (mushrooms) as living organisms.

Patterns in the Sky

The **Big** Idea

The positions of the sun, moon, and stars show patterns.

On this page, show what you learn as you read this chapter.

Essential Question

What are stars?

Essential Question

What causes day and night?

Essential Question

Why does the moon seem to change?

Essential Question

What causes the seasons?

Moon Journal

You need
- monthly calendar
- pencil
- crayons

What to Do

1. Go outside with a family member each night for one month.

2. Find the moon. Draw what you observe. Write the date on each picture.

3. After one month, review your drawings. Label the new, first quarter, full, and last quarter moon.

Draw Conclusions

1. Describe how the moon seemed to change.

_____

_____

_____

2. How many days are there between moon phases?

_____

Lesson 1

Finding Constellations

1. Scatter dried beans on a sheet of paper.

2. Look for shapes made by the beans. Draw them.

3. How is this like what people long ago did when they looked at the stars?

_____

_____

Lesson 2

Make a Sundial

1. Put the end of a stick in the ground, pointing straight up.

2. When it is 1:00, place one stone at the end of the stick's shadow. Place two stones at 2:00 and three stones at 3:00. Do not move the stones or the stick.

3. Use the stones and the stick's shadow to tell the time the next day.

4. Explain how your sundial works. _____

_____

Lesson 3

Model Moon Phases

1. Work with a partner. Wrap a ball in foil. Hold the ball while your partner shines a light onto it.

2. Slowly turn in place, keeping the ball in front of you. Observe how the light hits the ball.

3. What does your model show you about the moon phases? _____

_____

Lesson 4

Slanting Light

1. Shine a flashlight directly onto a sheet of paper.

2. Now tilt the paper so that the light hits it at a slant.

3. What looks different about the light on the page?

_____

_____

4. What does your model show you about the way sunlight hits Earth at different times of the year?

_____

_____

Vocabulary

 A **star** is a big ball of hot gases that give off light and heat. The sun is the closest star to Earth.

 A group of stars that form a pattern is a **constellation**. This constellation is called Orion.

 To **rotate** means to spin around like a top. Earth rotates one time every 24 hours.

The **moon** is a huge ball of rock that orbits Earth. The moon takes almost one month to go all the way around Earth.

A **season** is a time of year that has a certain kind of weather. The four seasons are spring, summer, fall, and winter.

What Are Stars?

The sun is a **star**. It is made of hot gases.
The hot gases give off light and heat.
The sun is the closest star to Earth.
We see other stars only at night.

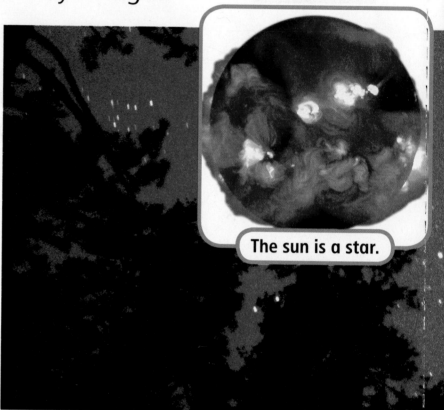

The sun is a star.

---

1. Underline the words that tell you what a star is made of.

2. How is the sun different from other stars?

   The sun is hot but
   the star is not.

6

Size

Some stars are smaller than the sun.
Some stars are larger than the sun.
Stars look small because they are far away.

stars

---

3. Are all stars the same size? _____No_____

4. Why do stars look small?

Its far away from you.

Brightness

Some stars look brighter than others.
They may be bigger or hotter.
They may be closer to Earth.

5. Why might one star look brighter than another star?

_____

_____

Patterns

A group of stars may form a **constellation**.
A constellation is a pattern of stars.

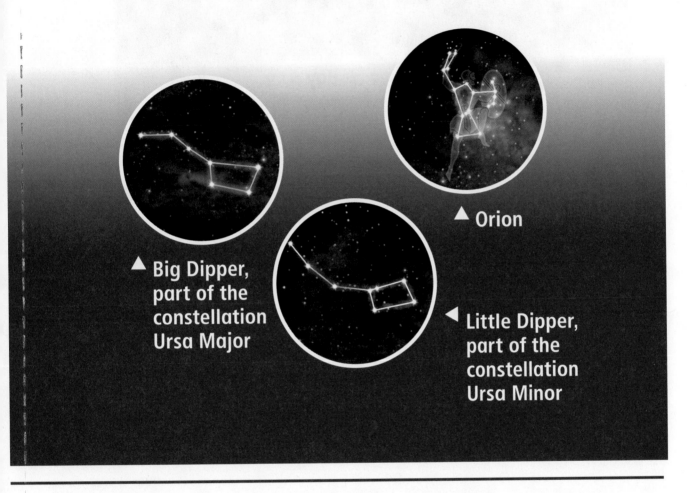

▲ Big Dipper,
part of the
constellation
Ursa Major

▲ Orion

◀ Little Dipper,
part of the
constellation
Ursa Minor

6. Circle the name of a constellation.

7. What is a constellation?

_____

_____

What Causes Day and Night?

day

The sun seems to move across the sky.
But the sun does not move. Earth moves.
Earth **rotates**, or spins like a top.
It takes about 24 hours to make one turn.

1. How long does it take Earth to rotate one time?

_____

2. Look at the picture. How can you tell which part of
Earth has day?

_____

_____

_____

At times, our part of Earth faces the sun.
That is when we have day.
Then, our part faces away from the sun.
That is when we have night.

---

3. Look at the picture. How can you tell which part of Earth has night?

_____

_____

_____

Shadows

morning

noon

Objects can block the sun's light.
This makes a shadow.
Sometimes shadows are short.
Sometimes shadows are long.

---

4. Draw an **X** on a shadow that is short.

5. How does a shadow form?

Objects can block the sun's light.

afternoon

As Earth rotates, the sun seems to move.
The sun's light shines on objects.
The direction of the light changes as the
day goes on.
This causes the shadows to change.

6. Which pictures show long shadows?

after noon

7. Why is the shadow different in each picture?

The direction of the light
changes as the day goes on.

Why Does the Moon Seem to Change?

Earth

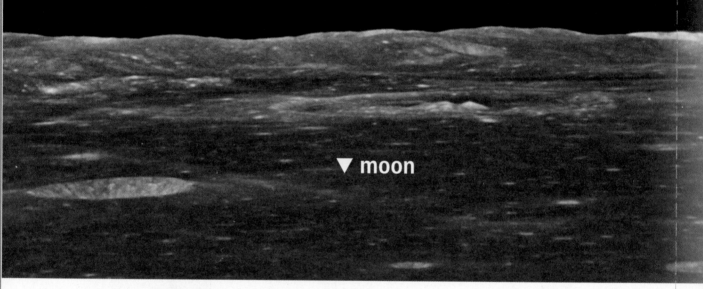

▼ moon

The **moon** is a ball of rock.
It moves in an orbit around Earth.
One full orbit takes about a month.

1. What is the moon made of?

   The moon is a ball of rock.

2. How long does it take for the moon to orbit Earth
   one time?

   It take for the moon
   to orbit about a month.

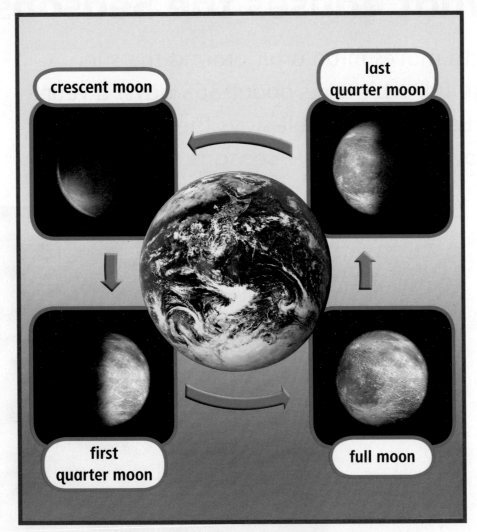

The moon reflects light from the sun.
The part of the moon we see changes.
That is why its shape seems to change.

3. Number the pictures to show the order
   in which the moon phases happen.

_____

_____

_____

_____

What Causes the Seasons?

Earth moves in an orbit around the sun.
One full orbit takes about 365 days, a year.
As Earth orbits, its tilt stays the same.
That is why we have **seasons**.

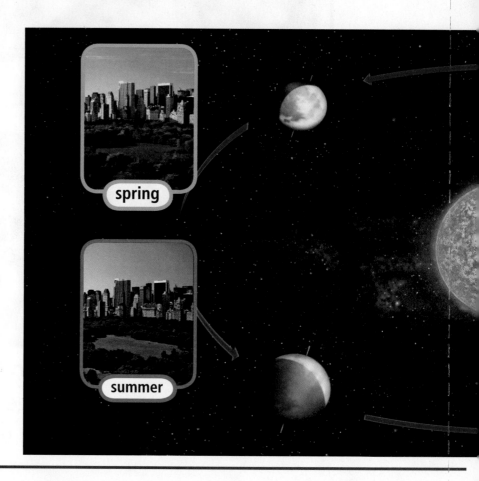

1. About how long does it take for Earth to orbit the sun one time?

_____

2. Circle the picture of the season where the top part of Earth is MOST tilted toward the sun.

At times, our part of Earth is tilted toward the sun.
That is when we have summer.
Then, our part is tilted away from the sun.
That is when we have winter.

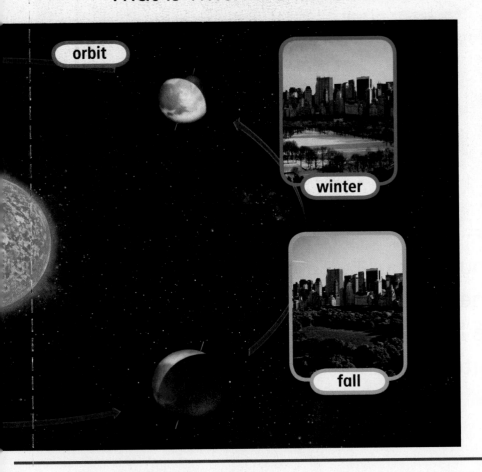

3. Circle the picture of the season where the top part of Earth is LEAST tilted toward the sun.

4. Draw a picture of your part of Earth in winter.

Seasons in Georgia

Which day has the most hours of daylight? Why?

Hours of Daylight in Georgia

day	number of hours of daylight
March 21	12
June 21	14
September 21	12
December 21	10

0 1 2 3 4 5 6 7 8 9 10 11 12 13 14 15

number of hours of daylight

Sunlight hits your part of Earth directly in summer.

The air is warmer.

There are more hours of daylight.

The days are often hot and sunny.

---

5. Draw a picture of your part of Earth in summer.

18

In Georgia, it can get very cold in winter. Sometimes it snows.

Sunlight hits your part of Earth at a slant in winter.
The air is cooler.
There are fewer hours of daylight.

6. Look at the graph. Are there more hours of daylight in summer or in winter?

_____

7. What season is it in the picture? How do you know?

_____

_____

_____

CRCT Practice

Fill in the circle in front of the letter of the best choice.

1. The sun is a star. Which is TRUE about the sun?

   ○ A. It is made of hot gases.
   ○ B. It is made of cold gases.
   ○ C. It is the biggest star.

   S2E1a

2. Look at the shadow in the picture. What time of day is it?

   ○ A. early morning
   ○ B. noon
   ○ C. late afternoon

   S2E2a

3. What season is it when the part of Earth where you live is tilted toward the sun?

   ○ A. summer
   ○ B. fall
   ○ C. winter

   S2E2c

4. **What can be measured with a shadow stick?**

   O A. how shadows change
   O B. how shadows grow
   O C. how the moon orbits Earth   **S2E2b**

5. **Look at the picture. What are patterns in the sky like this called?**

   O A. orbits
   O B. rotations
   O C. constellations   **S2E1a**

6. **Which is a picture of a full moon?**

○ A.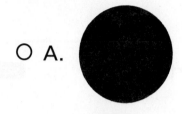

○ B.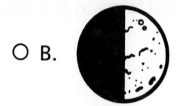

○ C.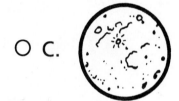

S2E2d

7. **Which season has days with the least amount of daylight?**

○ A. spring
○ B. fall
○ C. winter

S2E2c

8. **At what time of day are shadows the longest?**

    ○ A. morning

    ○ B. noon

    ○ C. night   `S2E2b`

9. **What causes day and night?**

    ○ A. the moon's orbit

    ○ B. Earth's rotation

    ○ C. the sun's rotation   `S2E2a`

10. **About how often does the pattern of moon phases repeat?**

    ○ A. every day

    ○ B. every 14 days

    ○ C. every 29 days   `S2E2d`

CHAPTER 3

Changes to the Earth

The Big Idea

Weather, plants, animals, and people can cause changes to the land.

On this page, show what you learn as you read this chapter.

Essential Question

How does weather cause changes?

Essential Question

How do living things cause changes?

Go online ▸ Student eBook
www.hspscience.com

What Happens to Trash?

What to Do

1. Bury the trash objects in the soil. Put the pan of soil in a sunny place.

2. Water the soil every three days. After two weeks, dig up the things. What do you observe?

You need
- trash objects (suggested items: lettuce, napkin, piece of foam cup)
- pan of soil
- water
- gloves

_____

_____

_____

Draw Conclusions

1. In what different ways do objects change?

_____

_____

2. How could trash harm the land?

_____

_____

Insta-Lab

Lesson 1

Flooding

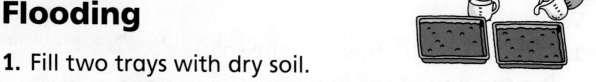

1. Fill two trays with dry soil.

2. Slowly pour 1 cup of water onto one tray.

3. Quickly pour 1 cup of water onto the other tray.

4. What happens?

_____

_____

_____

5. Why might some rains cause floods while others do not?

_____

_____

_____

26

Model an Oil Spill

1. Put some water in a jar, and add a little oil.

2. Dip a feather into the oily water. Then feel the feather.

3. Describe how the feather feels.

_____

_____

4. How do you think oil spills harm birds?

_____

_____

Vocabulary

 A **flood** happens when rivers and streams get too full. The water flows onto land.

 A **drought** is a long time with little rain that causes the land to get very dry.

 **Erosion** is when moving water changes the land by carrying rocks and soil to new places.

An **environment** is all the living and nonliving things in a place.

Waste that harms the air, water, or land is **pollution**.

How Does Weather Cause Changes?

flood

Rain may make a river too full.
Some water runs onto the land.
It is a **flood**.
Flood water may carry away soil.

1. Tell why a flood happens.

_____

_____

2. Circle the sentence that tells the effect of a flood.

drought

It may not rain for a long time.
Streams and ponds dry up.
The land gets very dry.
It is a **drought**.
Wind may blow some soil away.

---

3. Tell why a drought happens.

_____

_____

4. What can happen to soil during a drought?

_____

Fire Makes Changes

Lightning may hit trees in the forest.
The trees may burn.
Animals must move to safer places.

5. How can lightning start a fire?

_____

6. How does a fire affect animals?

_____

_____

Lightning can start a fire.

New plants grow after the fire.
Many animals come back.

_____

7. Tell what happens after a fire.

_____

_____

Water Makes Changes

Moving water changes the land.
It carries rocks and soil to new places.
This is called **erosion**.

8. Tell how erosion changes the land.

_____

_____

How Do Living Things Cause Changes?

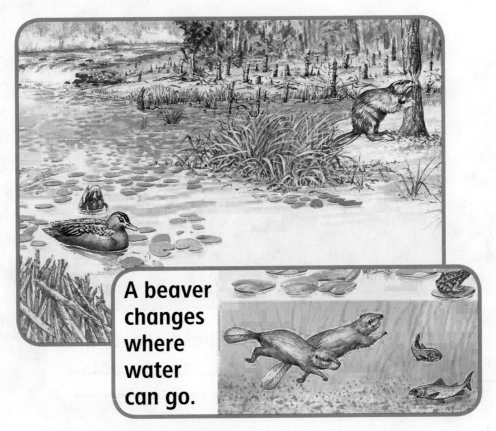

A beaver changes where water can go.

An **environment** has living things.
An environment has non living things.
Animals change their environments.

---

1. What kinds of things are part of an environment?

_____

2. Circle the sentence that tells how a beaver changes its environment.

Plants Make Changes

These plants are called sea oats.

Plants help their environments.
Plant roots hold soil in place.
This helps stop erosion.

3. What is one way that plants help their environments?

_____

This plant grows fast.

Plants can harm their environments.
Plant roots may grow in sidewalk cracks.
The plants break up the sidewalk.
Some plants grow over others.
The other plants do not get enough light.

4. What is one way that plants can harm
   their environments?

_____

5. A plant grows very fast. It grows over
   other plants. The other plants die.
   Tell why.

_____

People Make Changes

pollution

People make **pollution**.
It is waste that harms air, water, or land.
Smoke from cars makes the air dirty.
It can harm plants, animals, and people.

6. Why do you think it might be good for people to take a train instead of driving in cars?

_____

_____

_____

Pollution runs into the water.

Oil and waste make the water dirty.
Dirty water can make living things sick.

7. Does pollution help or harm the environment?

_____

8. What can make water dirty?

_____

Litter is trash people did not put into trash cans.

Litter makes the land dirty.
Plants get covered by litter.
They can not get the light they need.
Animals may eat litter and get sick.

---

9. What is litter?

_____

10. How does litter harm animals?

_____

Put litter in trash cans.
It will go to a landfill.
Then it will not pollute water and land.

---

11. Circle the picture that does NOT show a way people harm the environment.

People Waste Resources

People can waste resources.
They may want to use the land.
They cut down all the trees.
The trees may be wasted.

12. How do you think people use the wood from trees so that it is not wasted?

_____

_____

People may cut down trees to build houses.
Animals use the trees for homes and for food.
The animals must find new homes and food.
If they do not, they will die.

---

13. Tell one reason why people cut down trees.

_____

14. How does cutting down trees harm animals?

_____

_____

CRCT Practice

Fill in the circle in front of the letter of the best choice.

1. Which is one way that people can help the environment during a drought?

   ○ A. use very little water
   ○ B. use a lot of water
   ○ C. move to a safe, dry place   S2E3a

2. How might weather cause a fire?

   ○ A. People leave a campfire burning.
   ○ B. Gas from a car catches fire.
   ○ C. Lightning from a storm hits a tree.   S2E3a

3. Which is NOT a form of pollution?

   ○ A. trash
   ○ B. floods
   ○ C. oil spills   S2E3a

4. Look at the picture. What is one way the birds could cause a change to the environment?

○ A. fly

○ B. build a dam

○ C. build a nest

S2E3a

5. Where Rudy lives, it has not rained for a very long time. Many plants and animals have died. What is this time called?

○ A. a blizzard

○ B. a drought

○ C. a flood

S2E3a

6. What could smoke from this factory cause?

- ○ A. litter
- ○ B. air pollution
- ○ C. water pollution

S2E3a

7. What change takes the LEAST amount of time?

- ○ A. drought
- ○ B. erosion
- ○ C. flood

S2E3a

8. **Which makes air pollution?**

O A. bicycles

O B. plants

O C. trucks S2E3a

9. **Which of these is an effect of plant growth?**

O A. more air pollution

O B. less erosion

O C. more floods S2E3a

10. **How do beavers change their environments?**

O A. by building dams

O B. by growing trees

O C. by digging holes S2E3a

States of Matter

The Big Idea

Matter has different states. Matter can be a solid, liquid, or gas.

On this page, show what you learn as you read this chapter.

Essential Question

What is matter?

Essential Question

What are solids?

Essential Question

What are liquids?

Essential Question

What are gases?

GO online
Student eBook
www.hspscience.com

Hot Air

You need
- plastic bottle
- balloon
- bowl
- hot water

What to Do

1. Stretch a balloon over the top of an empty bottle.

2. Place the bottle in a bowl. Watch an adult pour hot water into the bowl. **CAUTION:** Be careful near the hot water!

3. Wait a few minutes. What happens?

_____

_____

Draw Conclusions

1. Why did the balloon fill with air?

_____

_____

2. What do you think will happen when the water cools?

_____

_____

Lesson 1

Matter and Space

1. Choose 3 classroom objects.

2. Arrange them in order from the lightest to the heaviest.

3. Which object has the most matter?

_____

_____

Lesson 2

Sort Classroom Objects

1. With a partner, gather some solid objects in your classroom.

2. Take turns describing their properties.

3. Sort the objects.

4. What property did you use to sort the objects?

_____

_____

Measure Two Ways

1. Use a measuring cup to find out about how many ounces equal 150 milliliters.

2. How did you find the amount?

_____

_____

3. What did you find out?

_____

Observe Air

1. Tightly pack a paper towel into the bottom of a cup.

2. Turn the cup upside down. Push the cup straight down into a bowl of water. Then pull it straight out.

3. What happens to the paper towel? Why?

_____

_____

_____

Vocabulary

All things are made of **matter**. Matter can be a solid, a liquid, or a gas.

Mass is the amount of matter in an object. Mass can be measured using a balance.

A **property** is one part of what something is like. Color, size, and shape are each a property.

A **solid** is the only form of matter that has its own shape.

Texture is the way something feels when you touch it. Sand has a rough texture.

A **centimeter** is a unit used to measure how long a solid is. Centimeters are marked on many rulers.

A **liquid** is a form of matter that takes the shape of its container.

Volume is the amount of space something takes up. Volume can be measured with a measuring cup.

A **milliliter** is a unit used to measure the volume of a liquid. Milliliters are marked on many measuring cups.

The balloons are filled with **gas**. A gas is the only form of matter that always fills all the space of its container.

What Is Matter?

Matter is what all things are made of.
All matter takes up space.
All matter has **mass**.
Mass is the amount of matter in an object.

---

1. Draw a circle around a kind of matter in the picture.

2. What is mass?

_____

Size is a **property** of matter.
Color and shape are other properties.
Properties can help you tell objects apart.

---

3. Circle words that tell properties of matter.

4. Use a property of matter to describe the shovel.

_____

What Are Solids?

A **solid** is a form of matter.
Some solids are hard. Some are soft.
Solids have different textures.
Texture is the way something feels.

---

1. Draw pictures to show three solids on this page.

2. What is texture?

_____

All solids have a shape of their own.
You can do things to change the shape.
You can bend a solid or break it or cut it.
But the shape does not change on its own.

---

3. Does solid matter change shape on its own?

_____

4. What is one way you can change the shape of a solid?

_____

Measuring Solids

balance

You can measure the mass of a solid.
A balance helps you measure mass.

---

5. Underline the word that tells what tool measures mass.

6. Circle the solid object the boy is measuring.

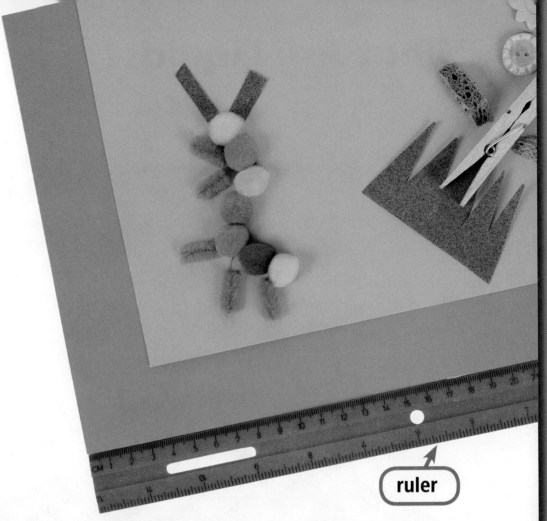

ruler

You can measure the size of a solid.
A ruler helps you measure size.
A ruler may measure in **centimeters**.
Or, it may measure in inches.

---

7. Circle the word that tells what tool measures size.

8. What are two ways a ruler may measure?

_____

What Are Liquids?

A **liquid** is a form of matter.
Liquids do not have their own shape.
They take the shape of their container.

---

1. What are the names of three liquids?

_____

2. Do liquids have their own shape?

_____

measuring cup

You can measure the **volume** of a liquid.
A measuring cup measures volume.
It may measure volume in **milliliters**.
Or, it may measure volume in ounces.

3. Circle the object that measures volume.

4. About how many ounces full is the measuring cup?

What Are Gases?

A **gas** is a form of matter.
A gas has no shape of its own.
A gas fills its entire container.
A gas has mass.

1. What is inside the bubbles?

_____

2. Turn back to page 60. What is one way liquids and gases are the same?

_____

Air is made up of gases.
You can not see or smell air.
But you can see what air does.

3. What makes the pinwheel turn?

4. What is one other thing air does in the picture?

Fill in the circle in front of the letter of the best choice.

1. **What are all things made of?**

   ○ A. liquid

   ○ B. gas

   ○ C. matter   **S2P1a**

2. **A solid is the only form of matter that has its own**

   ○ A. shape.

   ○ B. mass.

   ○ C. properties.   **S2P1a**

3. **Which forms of water are shown in these glasses?**

   ○ A. solid and gas

   ○ B. liquid and solid

   ○ C. gas and liquid   **S2P1a**

4. **Justin blew up this balloon. What state of matter is inside the balloon?**

  ○ A. solid
  ○ B. liquid
  ○ C. gas

S2P1a

5. **Texture is the way something feels when you**

  ○ A. taste it.
  ○ B. touch it.
  ○ C. smell it.

S2P1a

6. In this picture, which thing is a liquid?

- ○ A. glass
- ○ B. chalk
- ○ C. water

7. Which form of matter do the pictures show?

- ○ A. gas
- ○ B. solid
- ○ C. liquid

8. **How are solids different from liquids?**

   ○ A. Solids have more mass than liquids.

   ○ B. Solids take the shape of their containers.

   ○ C. Solids keep their own shape.   `S2P1a`

9. **Which can you usually NOT see?**

   ○ A. a gas

   ○ B. a liquid

   ○ C. a property   `S2P1a`

10. **Greg went for a walk. He found a purple stone. He breathed the air. He stepped in a puddle. Which was solid matter?**

    ○ A. the air

    ○ B. the puddle

    ○ C. the purple stone   `S2P1a`

CHAPTER 5

Changes in Matter

The Big Idea

Matter changes if it tears, dissolves, freezes, melts, burns, or cooks.

On this page, show what you learn as you read this chapter.

Essential Question

How can matter change?

Essential Question

How can water change?

Essential Question

What are other changes to matter?

GO online Student eBook
www.hspscience.com

Mixing Matter

What to Do

You need
- $\frac{1}{2}$ cup warm water
- spoon
- salt
- pie pan

1. Put a spoonful of salt into the warm water. Stir well.

2. Pour the mixture of salt and water into the pan.

3. Put the pan in a warm spot. Predict what will happen.

_____

Draw Conclusions

1. What happened to the water?

_____

2. What was left in the pan?

_____

3. What does this show about mixtures?

_____

_____

Lesson 1

Build and Measure

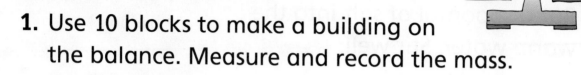

1. Use 10 blocks to make a building on the balance. Measure and record the mass.

_____

2. Then take the building apart. Measure and record the mass of the blocks.

_____

3. Did the mass change or stay the same?

_____

Lesson 2

Which Melts Faster?

1. Set out two plates. Put two ice cubes on each plate.

2. Place a lamp over one of the plates. Wait five minutes.

3. Which ice cubes melted more? Why?

_____

_____

Uncooked and Cooked

1. Draw pictures to show what spaghetti looks like before and after it is cooked.

Before

After

2. Describe how the spaghetti changes.

_____

_____

Vocabulary

The fruit salad is a **mixture** of different kinds of fruit. Kinds of matter in a mixture do not change into other kinds of matter.

To **dissolve** is to completely mix a solid with a liquid.

Evaporation is the change of water from a liquid to a gas.

Water in the form of a gas is **water vapor**.

Condensation happens when heat is taken away from water vapor. Water changes from a gas to a liquid.

The trees in this forest are **burning**. Burning is the change of a substance into ashes and smoke.

How Can Matter Change?

You can make mixtures with solids.
You can take a **mixture** apart.
The parts stay the same.

The fruit in this salad does not change size or shape.

---

1. Which state of matter is the fruit salad?

_____

2. Give another example of a mixture.

_____

You can mix some solids and liquids.
Sugar **dissolves** into water and lemon juice.
You can taste the sugar in the lemonade.

3. Underline the matter that is used to make lemonade.

4. What state of matter is the lemonade?

_____

Changing Matter

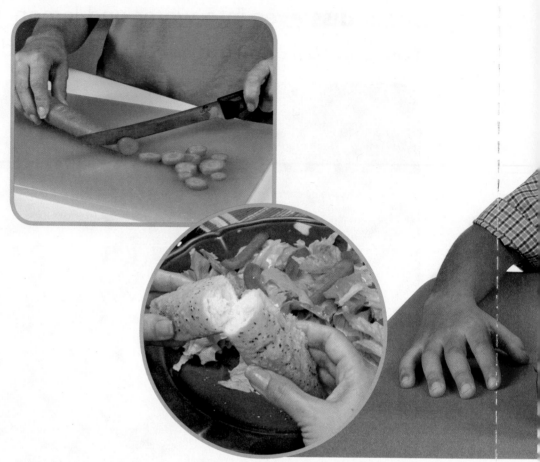

You can change the size of matter.
You can also change the shape of matter.
The mass of the matter does not change.

---

5. Draw one way that you can change matter.

Cheese is matter.
You can cut cheese into pieces.
The mass of the cheese stays the same.

6. What happened to the mass of the cheese when it was cut into pieces?

_____

7. How do you know?

_____

How Can Water Change?

Liquids may freeze and become solids.
This happens when heat is taken away.

1. Underline the temperature word for making water solid.

2. Circle the temperature word on page 79 for making water liquid.

Solids, like this ice pop, may become liquid.
This happens when heat is added.
When enough heat is added, the ice melts.

3. Draw lines to match up the kinds of matter.

frozen water   liquid

melted ice   solid

Evaporating and Boiling

Water can change from a liquid to a gas.
This happens when heat makes water boil.
When water boils, **evaporation** happens.
Water becomes **water vapor**, a gas.

---

4. Finish the sentence. Use the words **liquid** and **gas**.
When water boils, it changes from a _____
to a _____.

5. What is water in gas form called?

_____

Condensing

Liquid water forms on the outside of the glass.

Water vapor can change into liquid water.
This happens when heat is taken away.
Cold water takes heat from the air.
Condensation takes place on the glass.

---

6. Finish the sentence. Use the words **liquid** and **gas**.
When water condenses, it changes from a _____
to a _____.

7. Circle the example of condensation in the picture.

What Are Other Changes to Matter?

Burning changes matter into new matter.
It changes wood into ashes and smoke.
They can not change back into wood.

Cooking changes matter into new matter.
Heat can turn a marshmallow brown.
It can change meats and vegetables.
These foods can not change back again.

---

1. Circle something in the picture that is burning.

2. Draw an X on something in the picture that is cooking.

3. Turn back to page 74. How is mixing different from cooking?

_____

_____

_____

CRCT Practice

Fill in the circle in front of the letter of the best choice.

1. **How is the water changing in this picture?**

   ○ A. from solid to gas
   ○ B. from solid to liquid
   ○ C. from gas to liquid   S2P1b

2. **Which is NOT a way to change matter?**

   ○ A. slice it
   ○ B. mix it
   ○ C. observe it   S2P1b

3. **What happens to things in a mixture?**

   ○ A. They are gone.
   ○ B. They change into new matter.
   ○ C. They do not become new things.   S2P1b

4. **What is it called when water vapor changes from a gas to a liquid?**

   O A. condensation

   O B. melting

   O C. evaporation

5. **Look at the picture. What kind of change is happening to the matter?**

   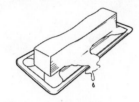

   O A. tearing

   O B. dissolving

   O C. melting

6. Which picture is NOT a mixture?

○ A.

○ B.

○ C.

S2P1b

7. What happens to water when it boils?

○ A. It evaporates.

○ B. It melts.

○ C. It condenses.

S2P1b

8. **How can you change a liquid to a solid?**

   ○ A. burn it
   ○ B. mix it
   ○ C. freeze it   `S2P1a`

9. **What happens when paper burns?**

   ○ A. Its matter changes.
   ○ B. Its shape stays the same.
   ○ C. Its color stays the same.   `S2P1b`

10. **Alan places a pan of water in the freezer for two hours. When he takes the pan out, the water is frozen. If he repeats this experiment, what is likely to happen?**

   ○ A. The water will boil.
   ○ B. The water will freeze.
   ○ C. The water will evaporate.   `S2P1b`

Energy

The Big Idea Energy causes matter to move or change. Light, heat, sound, and motion are some forms of energy.

On this page, show what you learn as you read this chapter.

Essential Question

What is energy?

Essential Question

What is light?

Essential Question

What is heat?

GO online Student eBook
www.hspscience.com

Heat on the Move

What to Do

You need
- bowl of hot water
- metal, plastic, and wooden spoons

1. Wait while an adult fills a bowl with hot water. **CAUTION:** Be careful near hot water!

2. Place each kind of spoon in the hot water.

3. Wait two minutes. Then carefully feel each spoon and compare the handles of the spoons.

Draw Conclusions

1. What can you tell about how heat moved?

_____

2. How does heat move differently through different materials?

_____

Water Power

1. Place three small objects in the middle of a pan.

2. Pour some water into the pan at one end.

3. What happens to the objects?

_____

_____

What Does Light Shine Through?

1. Shine a flashlight onto different materials. You might try wax paper, plastic wrap, newspaper, and construction paper.

2. Which ones does light pass through?

_____

_____

_____

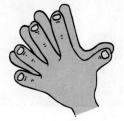

Making Heat

1. How do your hands feel? Are they warm or cool?

_____

2. Rub your hands together for 20 seconds.

3. Tell how your hands feel now and how they changed.

_____

4. What force did you use to make your hands feel warmer?

_____

Vocabulary

 **Energy** is something that can cause matter to move or change. Heat, light, sound, and motion are forms of energy.

 The sun and fire give off **light** energy. Light is a form of energy that lets you see.

 **Heat** is a form of energy that makes things warmer. Heat can be used to cook food.

 **Sound** is energy you can hear. The instruments make different sounds.

 Energy from the sun is **solar energy**.

When something moves, it is in **motion**. The sled is in motion.

To **reflect** is to bounce off. Light reflects when it hits most objects.

Friction is a force that slows down objects when they rub against each other. Friction also causes the objects to get warmer.

Electricity is a form of energy. People make electricity by using other kinds of energy.

A **thermometer** is a tool that measures an object's **temperature**. Temperature is how hot or cold the object is.

What Is Energy?

Energy can cause matter to move or change.
Light is a form of energy.
Heat is also a form of energy.
Sound is a form of energy.

---

1. Name two forms of energy.

Sound and light.

2. Draw an example of sound energy.

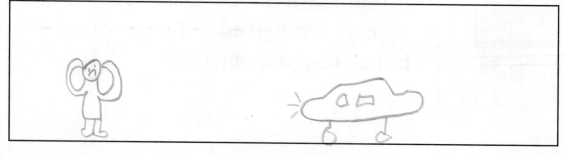

Where Energy Comes From

moving water

sun

Solar energy is energy from the sun.
Some energy comes from wind.
Wind can cause objects to be in motion.
Energy also comes from moving water.
Energy comes from fuels like coal and oil.

---

3. Circle the picture that shows where solar energy comes from.

4. Where else can energy come from?

<u>Solar energy.</u>

What Is Light?

Light is a form of energy that lets you see.
Light travels in straight lines.
When light hits objects, it is reflected.
You see things because objects **reflect** light.

---

1. How does light travel?

In straight lines.

2. What happens when light hits objects?

It is reflected.

Shadows

Light can pass through some things.
It can not pass through other things.
An object that blocks light makes a shadow.

---

3. What happens when an object blocks light?

It make shadow

4. Circle a shadow in the picture above.

What Is Heat?

Heat makes things warmer.
Fire and fuels give off heat as they burn.
Oil and wood are fuels. Natural gas is a fuel.
People use fuels to keep warm and to cook.

---

1. How do people use fuels?

It give off heat as they burn.

2. What type of energy do fuels give off as they burn?

Natural gas.

Friction

Ríction

When objects rub together, they get warm.
The heat is caused by **friction**.
Friction slows down the objects.

---

3. Underline the word that explains why objects get warm when they rub together.

Heat and Electricity

Rich.N

nuclear
energy
station

---

4. Use numbers to show the order in
which these sentences happen.

Heat changes water into steam.   1

Machines make electricity.   3

Steam turns machines.   2

Heat is used to make electricity.
First, heat changes water into steam.
The steam turns machines.
The machines make electricity.

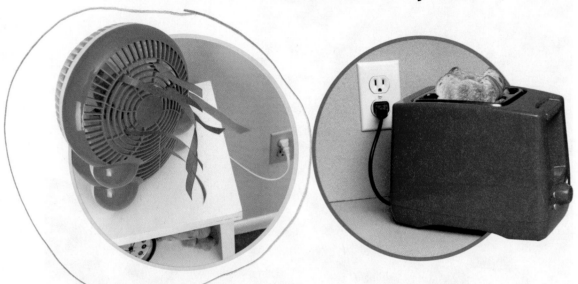

Power lines carry electricity to buildings.
Electric wires go to outlets in walls.
Some things change electricity to heat.
Other things may change it to light or motion.

5. Give two examples of objects in your house that use electricity.

Oven

phone

6. Circle the object that is changing electricity to heat energy.

Heat Travels

Heat moves from warmer to cooler things.
Heat travels easily through things like metal.
It does not travel easily through plastic.
It does not move easily through oven mitts.

---

7. How does heat move?

Heat travels easily through thing like metal.

8. Why should people use oven mitts to take things out of the oven?

When you put something in oven and when you take it out, it really hot.

Measuring Heat

Temperature tells how hot or cold a thing is.
A **thermometer** measures temperature.
It may measure in degrees Fahrenheit.
Or, it may measure in degrees Celsius.

---

9. Circle the thermometer that shows a hotter temperature.

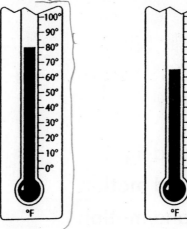

CRCT Practice

Fill in the circle in front of the letter of the best choice.

1. **What kind of energy is being used to toast the marshmallow?**

   ○ A. light energy
   ○ B. heat energy
   ○ C. sound energy   **S2P2b**

2. **What is energy from the sun called?**

   ○ A. solar energy
   ○ B. energy of motion
   ○ C. wind energy   **S2P2a**

3. **What types of energy does the lamp give off?**

   ○ A. heat and light
   ○ B. light and motion
   ○ C. heat and motion   **S2P2a**

4. **Which of these is a use of energy?**

   O A. falling rain
   O B. a moving bus
   O C. changing seasons   S2P2b

5. **What tool is used to measure how much heat an object has?**

   O A. balance
   O B. ruler
   O C. thermometer   S2P2b

6. **What happens when an object blocks light?**

   O A. A rainbow is formed.
   O B. The object disappears.
   O C. A shadow is formed.   S2P2b

7. **Look at the pictures. Which spoon will get the hottest?**

wood   plastic   metal

○ A. the metal spoon
○ B. the wood spoon
○ C. the plastic spoon

S2P2b

8. **Which form of energy is shown in the picture?**

○ A. oil
○ B. wind
○ C. light

S2P1b

9. **Which form of energy does this lamp use to get power?**

○ A. batteries
○ B. moving water
○ C. electricity

S2P2a

10. **What is one way people use heat?**

○ A. to turn windmills
○ B. to boil water
○ C. to cool homes

S2P2b

Motion

The Big Idea

Objects can move in many ways. Pushes and pulls can change the speed and direction of objects.

On this page, show what you learn as you read this chapter.

Essential Question

What are ways things move?

Essential Question

What makes things move?

Ramps and Rolling

What to Do

1. Make a ramp using two books and a piece of wood.

2. Roll a toy car down the ramp. Measure how far it rolls after it reaches the bottom. Record your answer in the table.

You need
- books
- thin piece of wood
- toy car
- paper
- pencil

Distance Car Rolled	
ramp with 2 books	
ramp with 3 books	
ramp with 4 books	

3. Use books to make the ramp higher. Roll and measure again. Record your answers in the table above.

Draw Conclusions

Why does the car roll a different distance each time?

_____

_____

Insta-Lab

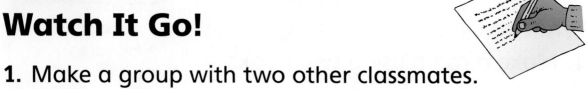

Lesson 1

Watch It Go!

1. Make a group with two other classmates.

2. Observe one person from your group moving at different speeds and in different directions for one minute.

3. Write sentences about what you observe.

_____

_____

4. Compare your observations. Did you observe the same thing?

_____

_____

Change Direction of a Ball

1. Show how you can change the direction of a ball by kicking it.

2. What happens when you kick the ball on one side?

_____

_____

3. What happens when you kick the ball harder?

_____

4. What conclusion can you draw about force?

_____

_____

Vocabulary

 **Motion** is a change of position. When something moves, it is in motion.

 **Direction** is the path a moving object takes.

 How fast an object moves or how far it moves in a certain amount of time is its **speed.**

A **force** is a push or a pull that can cause an object to move. A force can also change the way an object is moving.

Friction is a force that slows down objects when they rub against each other. Friction causes the objects to get warmer.

What Are Ways Things Move?

When you walk, you are in **motion**.
You move from one place to another.
There are many kinds of motion.
Some things move in a straight path.

1. Draw an **X** on something in the picture that is in motion.

2. What is one way things can move?

   A car.

Things can change **direction**.
The ball moves in a straight line.
If something touches it, its path changes.

3. Draw an example of how you could change the direction of a soccer ball.

Speed

Things may move fast or slow.
Speed is a measure of how fast something moves.

We can measure the speed of each rider.

---

4. Finish the sentence. Use the words **speed** and **fast**.

_____ is a measure of how _____ or slow something moves.

5. Name something that moves faster than a bike.

_____

Think about bikers in a race.
One biker moves faster than all the others.
The biker moves with the greatest speed.
That biker wins the race.

6. Draw an **X** on the biker with the greatest speed.

7. Circle the biker with the next greatest speed.

Changes in Motion and Speed

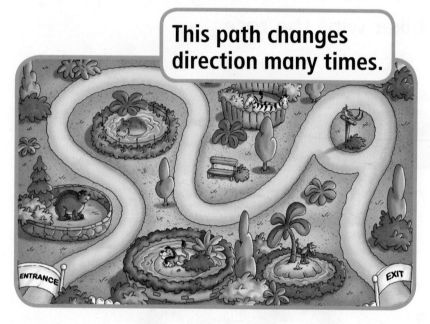

This path changes direction many times.

Things can move.
Things can speed up.
They can slow down.
They can change direction.
Things can stop or start.

Finish the sentences using the words in the box.

speed	direction

8. A path can change _____ many times.

9. Molly is at the entrance to the zoo. She wants to see the zebra. She can increase her _____ to get there faster.

The horse is speeding up.

A horse can move.
A horse might walk.
It might start running.
It changes speed.
It moves in many ways.

---

10. Tell two ways a dog can move.

_____

_____

What Makes Things Move?

pull   push

A **force** can make something move.
The force may be a push or a pull.
A pull moves something toward you.
A push moves something away from you.

1. Circle the words that tell about two kinds of forces.

2. What is the difference between a push and a pull?

_____

_____

A throw is a push.

A force can change the direction of a ball.
The girls are throwing the ball.
The throw, or push, is a force.

3. Draw a picture of you pushing something.

121

Strength of Forces

A force moves the ball.

It takes more force to move heavy things.
It takes less force to move light things.
A push may slow something down.
The girl uses a force to stop the ball.

4. Circle which would take more force.

rolling a bowling ball   **rolling a basketball**

5. You want to roller skate faster than your brother.
Do you need more or less force than your brother?

Friction

Friction is a force.
It can slow things down.
Friction happens when things rub together.
It is hard to ride a bike on grass.
The bumpy grass has a lot of friction.

---

6. Write a sentence about a surface that has very little friction.

_____

_____

_____

Fill in the circle in front of the letter of the best choice.

1. **What happens when you push a swing?**

   ○ A. It moves toward you.

   ○ B. It gets heavier.

   ○ C. It moves away from you.   S2P3a

2. **Bella kicks a ball. What force did Bella use to kick the ball?**

   ○ A. a push

   ○ B. a pull

   ○ C. direction   S2P3a

3. **What force makes it hard to push a chair across the floor?**

   ○ A. friction

   ○ B. motion

   ○ C. speed   S2P3a

4. **Dan kicks a light ball. Then he kicks a heavy ball with the same amount of force. Which ball will move farther?**

   ○ A. the light ball

   ○ B. the heavy ball

   ○ C. they will move the same distance   **S2P3b**

5. **How did the boy make the ball move?**

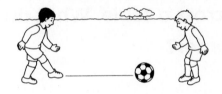

   ○ A. with a pull

   ○ B. with a push

   ○ C. in a curved path   **S2P3a**

6. **What is shown in the picture?**

   ○ A. a push
   ○ B. a pull
   ○ C. a change in direction   S2P3a

7. **You are pushing a book across a table. What do you change if you push harder?**

   ○ A. direction
   ○ B. motion
   ○ C. speed   S2P3b

8. **Which words can describe the speed of a moving object?**

   ○ A. curved, straight
   ○ B. fast, slow
   ○ C. far, near   S2P3b

9. **Which of these uses motion energy?**

   ○ A. box
   ○ B. picture frame
   ○ C. fan   S2P2b

10. **Kelly is walking. She is pulling a toy across the floor. What happens if she begins to run?**

   ○ A. The toy moves at the same speed.
   ○ B. The toy moves faster.
   ○ C. The toy moves slower.   S2P3b

Life Cycles

The Big Idea Animals, plants, and fungi are living things. Living things have life cycles that we can predict.

On this page, show what you learn as you read this chapter.

Essential Question

What are some animal life cycles?

Essential Question

What are some plant life cycles?

Essential Question

How do seasons change plants?

GO online

Student eBook
www.hspscience.com

Which Grow First?

What to Do

You need
- seeds of one kind of plant
- two cups
- soil
- water

1. Plant a few seeds in two cups filled with soil.

2. Make sure the seeds in both cups get water and sunlight.

3. Observe which seeds start to grow first.

Draw Conclusions

1. Did all the seeds start to grow at the same time?

   _____

   _____

2. Why do you think this happened?

   _____

   _____

3. What will happen as the plants continue to grow?

   _____

   _____

Insta-Lab

Sort It!

1. Compare the Picture Cards of cats, mice, frogs, butterflies, and dragonflies.

2. Sort them by animal. Sequence the stages of each animal's life cycle.

3. How are the life cycles alike and different?

_____

_____

Observe a Flower

1. Carefully take apart a flower.

2. Use a hand lens to observe the parts of the flower.

3. Draw a picture of what you observe.

How a Tree Changes

1. Look out a window. Draw a picture of a tree you see.

2. Draw a picture of what the tree will look like in a different season.

3. Write the names of each season below your pictures.

Vocabulary

All the stages, or times, of the life of an animal or plant are its **life cycle**.

This frog is a fully grown animal. It is an **adult**.

A **tadpole** is a young frog that hatches from an egg and lives in water.

A **larva** is the newly hatched form of some insects. A caterpillar is a larva.

A caterpillar becomes a **pupa** before it becomes a butterfly.

A **seed** is a plant part from which some new plants grow.

The **roots** of a plant hold the plant in the soil and take in water and nutrients. The **stem** of a plant holds up the plant.

Flowers are plant parts that make seeds. Part of each flower becomes a **fruit**. The fruit holds the seeds.

A **season** is a time of year that has a certain kind of weather.

The sweet liquid in flowers is called **nectar**.

What Are Some Animal Life Cycles?

2 kitten about 3 weeks old

1 adult cat and kittens

Every animal has a **life cycle**.
A life cycle is all the times of the animal's life.
This is a cat's life cycle.

---

1. Circle the words that tell what a life cycle is.

2. What is the last stage in a cat's life cycle?

The last stage of a cat is a adult cat

3 young cat about 6 months

4 adult cat

A kitten grows to be an **adult** cat.
The adult cat has its own young.
A new life cycle begins.

---

3. Tell about a cat's life cycle.

First the cat was

Life Cycle of a Frog

This is a frog's life cycle.
Mother frogs lay eggs.
Young frogs, or **tadpoles**, hatch from the eggs.

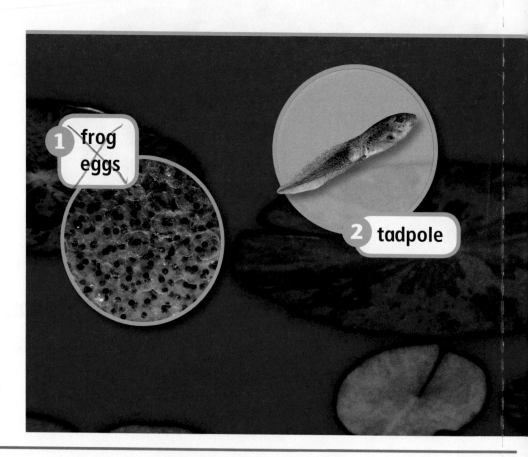

1 frog eggs

2 tadpole

4. What is the first stage in a frog's life cycle?

frog egg

5. Draw an **X** over the tadpole.

The tadpole grows and changes.
It becomes a fully grown frog.
Then it can have young.
A new life cycle begins.

3 growing tadpole

4 frog

6. How are tadpoles and frogs different?

Tadpole don't live on land
like frog.

Life Cycle of a Butterfly

This is a butterfly's life cycle.
A butterfly begins life as an egg.
A **larva**, or caterpillar, hatches from the egg.
The larva becomes a **pupa**.

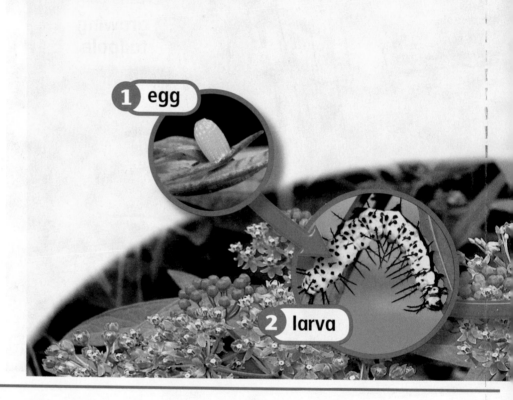

1 egg

2 larva

---

7. Circle the word that is another name for larva.

8. Put these stages in the order in which they happen.

pupa _____

butterfly _____

egg _____

The pupa makes a hard covering.
It changes inside the covering.
A butterfly comes out of the covering.
The butterfly can have its own young.

5 adult butterfly

4 butterfly coming out

3 pupa

9. Why do you think a pupa makes a hard covering?

_____

_____

_____

Life Cycle of a Bird

This is a bird's life cycle.
A mother bird lays eggs.
The mother or father keeps the eggs warm.
A chick grows inside each egg.
The chick breaks the egg to come out.

1 eggs in nest

2 chick just hatched

10. How does the chick get out of the egg?

_____

11. How does a chick get food?

_____

_____

The parents feed and take care of the chick.
The chick gets new feathers.
The adult bird can have its own young.

3 chick almost ready to fly

4 adult bird

12. Draw another animal that begins life as an egg.

What Are Some Plant Life Cycles?

bean plant starting to grow

Every plant has a life cycle.
A bean plant starts as a **seed**.
A tiny plant and some food are inside the seed.
The plant uses the food as it starts to grow.
First, **roots** grow. Roots take in water.

---

1. Circle the words that tell what is inside a seed.

2. Underline the words that tell what roots do.

bean plant growing from seed

Then, a **stem** grows up toward the light.
Leaves and more stems grow.
Last, the plant makes seeds.
The seeds may grow into new plants.

3. Draw a plant.
   Label its parts.
   Use these words.

   roots

   stem

   leaves

Flowers, Fruits, and Seeds

peach and peach seed

Some plants have **flowers**.
Part of each flower becomes a **fruit**.
The fruit grows around the seeds.
The fruit holds and protects the seeds.
A new life cycle begins.

4. Name two other plants that have flowers.

_____

5. How does the fruit help the seeds?

_____

Life Cycle of an Oak Tree

small tree growing from acorn

fully grown tree

Oak trees start as seeds in acorns.
A small plant grows from the acorn.
The plant grows into a big tree.
Flowers grow on the branches.
Each flower becomes an acorn.
A new tree can grow from each acorn.

6. Circle the name of an oak tree's seed.

7. Tell how a fully grown oak tree makes new oak trees.

_____

_____

_____

How Do Seasons Change Plants?

dogwood tree in spring

A **season** is a time of year.
Trees change when the seasons change.
Flowers grow on branches in spring.
Bees drink the sweet **nectar** in the flowers.
Green leaves cover the tree in the summer.
Green fruit grows on the branches.

1. Draw a picture
   of how a tree looks
   in spring.

dogwood tree in fall

The leaves and fruit turn color in fall.
The leaves fall off the tree.
Animals eat the fruit.
The branches are bare in winter.
Flowers grow on the branches again in spring.

2. In what season are the branches of some trees bare?

_____

3. Tell how a tree changes in fall.

_____

_____

Evergreen Trees

live oak tree

pine trees

Evergreens keep their leaves in winter.
The live oak is an evergreen in Georgia.
Pine trees are evergreens, too.

4. Circle the word that describes a tree that keeps its leaves in winter.

5. Draw how a pine tree looks in winter.

Wildflowers

black-eyed Susan

Georgia has many wildflowers.
Most wildflowers bloom in spring.
Some wildflowers die when it is cold out.
Georgia does not get very cold in fall.
Some wildflowers can still be seen.

6. Ricky went for a walk. He saw many wildflowers blooming. What was MOST LIKELY the season?

_____

How Crops Change

tomato plant

Tomatoes are crop plants.
In spring, tomato plants grow flowers.
Green tomatoes grow in summer.
They turn red in late summer and fall.
Pecans and onions are crop plants, too.

7. Circle the names of two crop plants.

8. Underline the season when tomato plants grow flowers.

Mushrooms

mushrooms

Never eat wild mushrooms!

Mushrooms are grown on special farms.
A mushroom is not a plant.
But it is a living thing.
It is a kind of fungus.

9. How are a mushroom and a plant alike?

_____

10. What kind of living thing is a mushroom?

_____

_____

Fill in the circle in front of the letter of the best choice.

1. **How does a bean plant begin its life cycle?**

   ○ A. as a seed

   ○ B. as a nut

   ○ C. as a root   **S2L1c**

2. **Look at the picture of the tree. Which season is MOST LIKELY beginning?**

   ○ A. summer

   ○ B. spring

   ○ C. fall   **S2L1b**

3. **What is another name for a butterfly larva?**

   ○ A. pupa

   ○ B. caterpillar

   ○ C. insect   **S2L1a**

4. **What is the living thing shown in the picture?**

  ○ A. a nut
  ○ B. a flower
  ○ C. a mushroom

5. **Look at the pictures of the trees throughout a year. What season is shown in the picture labeled 4?**

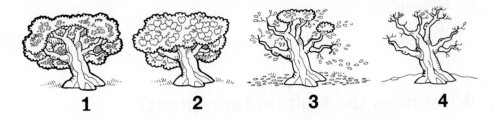

  ○ A. fall
  ○ B. winter
  ○ C. spring

6. **Look at the picture of the butterfly's life cycle. What do we call Stage 3?**

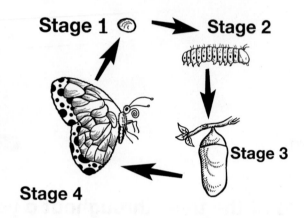

Stage 1   Stage 2

Stage 3

Stage 4

○ A. egg

○ B. larva

○ C. pupa

7. **What does the fruit do for a plant?**

○ A. holds and protects seeds

○ B. makes seeds

○ C. makes new plants

8. **Which kind of plant stays green in winter?**

○ A. a dogwood tree

○ B. a pine tree

○ C. a wildflower **S2L1b**

9. **How does an adult bird care for eggs in the nest?**

○ A. by cracking them open

○ B. by cleaning them

○ C. by keeping them warm **S2L1a**

10. **How are these two things the same?**

○ A. They are both plants.

○ B. They are both animals.

○ C. They are both living things. **S2L1d**